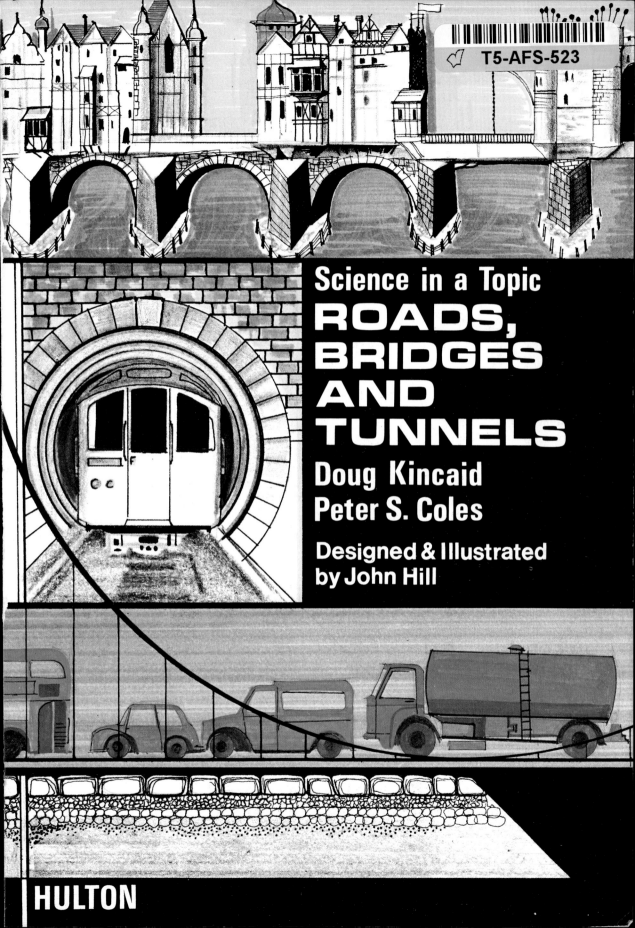

Science in a Topic
ROADS, BRIDGES AND TUNNELS

Doug Kincaid
Peter S. Coles

Designed & Illustrated
by John Hill

HULTON

ISBN 0 7175 0852 8
First published 1979
Hulton Educational Publications Ltd

SCIENCE IN A TOPIC
ROADS, BRIDGES AND TUNNELS

About this book

This book is different from most others because:

1. It is not complete, but only part of a study — the science part. There will be a need to use many other books to find out about other aspects of the topic — history, geography . . .

2. It will not tell you information but will only ask you questions and suggest ways that you might find the answers for yourself. Many of the suggestions were some children's ways of trying to find an answer — you may have better ideas.

3. It is hoped that arising from these questions other questions will occur to you — do pursue these. (Your own questions and the ways you find to answer them are really the most important.)

4. You do not need to work through the book in the order set out; the sections of work can be done in the order that you wish.

5. There is no need to complete all of one section. If the work becomes harder as you progress through a section, see how far you can go.

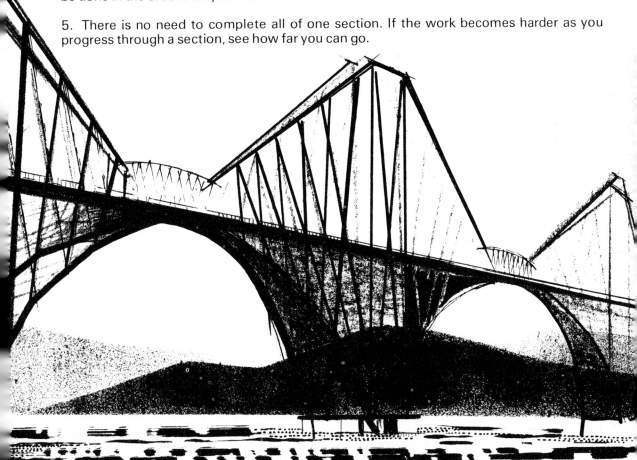

CONTENTS

SCIENCE IN A TOPIC SERIES

by Doug Kincaid, County Staff Advisory Teacher, Science, Buckinghamshire.
  Peter S. Coles, B.Sc., Chief Adviser, Berkshire.

Other Titles:
Ships
Houses and Homes
Clothes and Costume
Communication
Food
Moving on Land

Published by Hulton Educational Publications Ltd., Raans Road, Amersham, Bucks.
Printed Offset Litho by Martin's of Berwick.

UNDER AND OVER

Road and rail are two life-lines of our world.

They carry our food, our goods and ourselves.

Sometimes there are barriers. Mountains and rivers have to be crossed.

Bridges span gorges and gaps.

Tunnels pass through mountains and under rivers.

SECTION ONE

AROUND THE WORLD

Here are some famous roads, bridges and tunnels.
Where are they?
What do they link, span or cross?
There is a clue with each picture.

The skyscraper skyline will help to 'fix' this city.

This road is called the M4.

This is the Simplon Tunnel.

You probably know a song in French about this bridge.

This is the bridge of the gold and silversmiths in a famous Italian city.

6

FLUVIUS

South Warke

Roads, bridges and tunnels have always been important.

Look in other books for pictures and information, to find more about the great stories of engineering.

Here are some ideas for research.

Why does the canal bridge have a steep, humped arch?

What did 'legging' have to do with the canal tunnel?

This is a toll-house. What happened here?

Bridges are often destroyed in wars. This is because they are such important links.

Find out about the Rhine Bridges in World War II.

London grew around the first bridge upstream across the Thames. What other cities can you find that grew around a bridge?

Books that will help are listed on the back cover.

THE RECORD HOLDERS

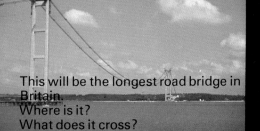

This will be the longest road bridge in Britain.
Where is it?
What does it cross?

This is one of the largest and most famous steel arch bridges in the world.
Where is it?

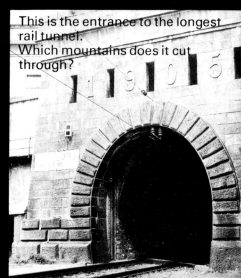

This is the entrance to the longest rail tunnel.
Which mountains does it cut through?

This is the longest span bridge built of brick in the world.
Where is it?
Who built it?

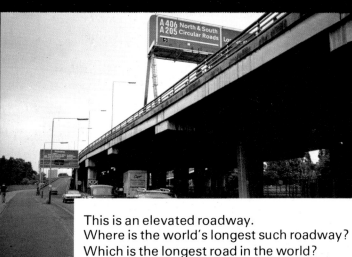

This is part of the longest tunnel of any underground railway.
Where is it?
How long is it?

This is an elevated roadway.
Where is the world's longest such roadway?
Which is the longest road in the world?

ROADS

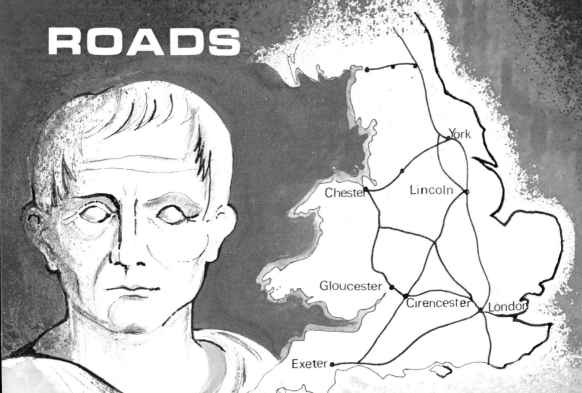

Good roads are vital to a modern nation. A good road network means people and goods can be moved with speed and ease.

The first roadways were made by people's and animals' feet. They were trackways made by use.

The roads built by the Romans are famous. They reached every corner of their empire.

For hundreds of years most roads were mere muddy tracks.

In the eighteenth century coaches needed much better roads. On pages 10–12 you can investigate the work of the great road-builders of this time.

Good roads can sometimes create problems.

ROAD BUILDING

The road engineer must find out what the soil is like. Some soils are unsuitable. This is important for planning the route of the road.

Before any building begins, soil samples are taken along the route.

These samples are taken to a laboratory and tested.

Test 1 – Compressibility
(This means how much the soil can be squashed and made firm.)

Test 2 – Cohesion
(This means how much the soil particles stick together.)

Test 3 – Permeability
(This means how much the soil lets water pass through it.)

Test 4 – Shrinkage
(This means by how much a sample of soil can become smaller.)

Collect some soil samples from around your school (and other places).

A soil auger is a good way to take samples.

A trowel can also be used.

Try some road engineering tests.

Bituminous surface

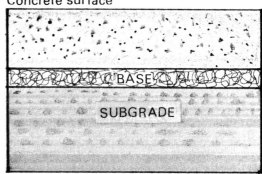

Concrete surface

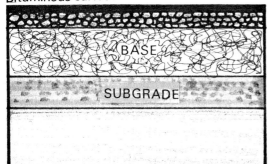

Test 1 – Compressibility

Fill a tin with your soil sample.
Use a rammer.
Pack down the soil as firmly as you can.
Measure the new level.

It is important that it does not compress more later, when the road is built. Test your sample again after a few days. Try wet and dry samples of the same soil.

Record:

Collect and try soils from other places (or use peat, sand and clay).

Soil sample	Height of soil in tin A	Height of compressed soil B	A–B
↓	↓	↓	↓

Test 2 – Cohesion

Roll your sample into a ball or sausage.
Does it stay in the shape, or crumble?
Does it need a little water to make it stick?
How much water? (Add a little at a time and measure.)

(If you are comparing soils, remember to make your test fair. Start with the same measure of soil. Add water measure by measure. Mix each thoroughly.)

Leave your samples in a good place to watch any changes.
Record these changes and how long they have taken.
What do you think caused them?
Try other places to check your ideas.
It is important to the road engineer for the soil to stay firm and stable.

11

ROAD BUILDING

Test 3 – Permeability

Set up this experiment.
Pour in 50ml of water.
Time how long the 50ml takes to drain through.
Measure how many ml have drained through
after 5 minutes, 10 minutes, 20 minutes . . .

Repeat the experiment with other soils.

Too much water in the soil under a road can cause
dangerous movement.

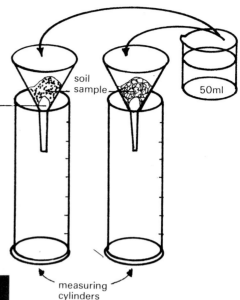

Test 4 – Shrinkage

Make a 20cm × 5cm slab of clay.
Leave it to dry.
Measure any changes.

Shrinkage under a road can be serious.

The road-builders lay a foundation for their
road.

The type and thickness will depend on their
discoveries from experiments like yours.

Which of your samples would be good for
road-building?

A good road surface keeps water away from
the foundations.

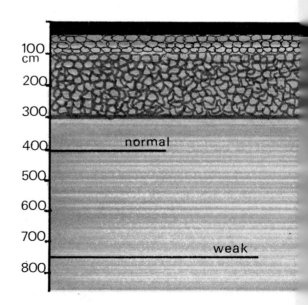

STRAIGHT ROADS

The job of building a new road or motorway is very complicated.

The first task is to plan the best path for the road.

The setting out of this route is done by a surveyor.

He uses an instrument like this to help him. It is called a theodolite. Now he also uses modern inventions including radio and lasers.

The Romans were famous as builders of straight roads.

To lay right-angle crossings they used a groma (shown below). You could make a model and try some surveying.

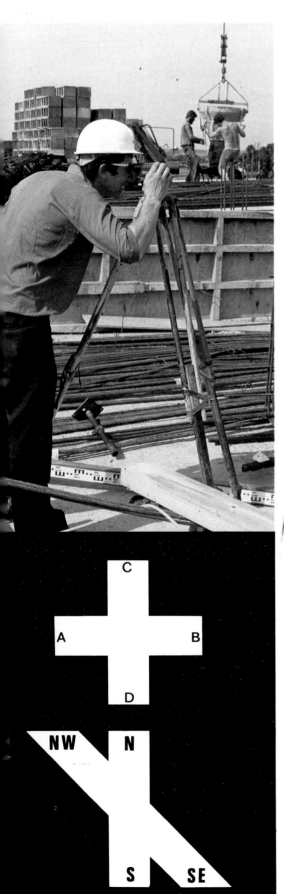

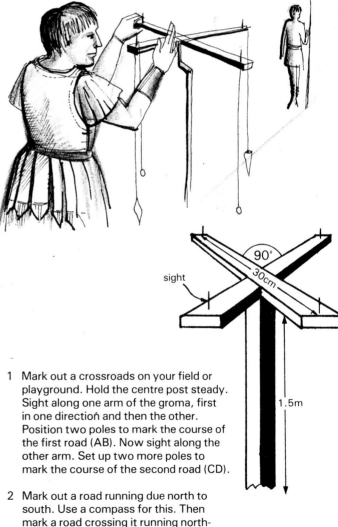

1 Mark out a crossroads on your field or playground. Hold the centre post steady. Sight along one arm of the groma, first in one direction and then the other. Position two poles to mark the course of the first road (AB). Now sight along the other arm. Set up two more poles to mark the course of the second road (CD).

2 Mark out a road running due north to south. Use a compass for this. Then mark a road crossing it running north-west to south-east.

CAMBERS AND CURVES

The last part of the road to be made is the surface.
This must prevent skidding.
The road needs to give a good grip to the tyres of the cars and lorries.
The road must be drained of water.

Generally the road surface is given a curve. The water will then drain off. This curve is called the camber.

How well do different slopes drain water?

Experiment – find out.

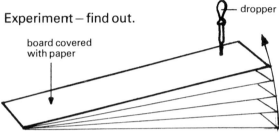

Investigate different slopes with a dropper.

You could also investigate different materials (plastic, roofing felt, sandpaper).

Most of the roads have curves and bends.
When there is a curve there are usually markings on the road.

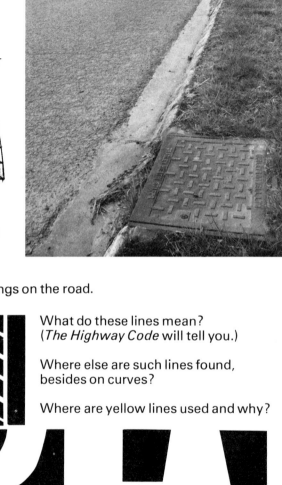

What do these lines mean?
(*The Highway Code* will tell you.)

Where else are such lines found, besides on curves?

Where are yellow lines used and why?

Imagine you are a road engineer.
What lines would you mark on these roads?
Make a large drawing and paint in your ideas.
Talk about your charts with other people.
Use *The Highway Code* to help you.

CAT'S EYES

Cat's eyes mark the centre and lanes of roads. Coloured cat's eyes mark motorway exits. They are a very simple but a very valuable invention.

Light from the headlamps of cars is reflected. Reflect means to 'send back' the light. The road lights up in front of the driver.

What other things can you find that will reflect light?
Here is a collection you could make.
Use a torch.
Sort them into things that reflect light and things that do not.

A bicycle has a reflector.

Try this with a torch.

Now try a dirty reflector.

How are the cat's eyes kept clean so that they will reflect well?
Look what happens when a car passes over.

| cat's eye before compression | tyre compresses cat's eye | reflectors wiped clean with the edge of rubber pad |

This important invention uses a very simple scientific idea. It must have saved thousands of lives. It is often the simple idea that makes the best invention.

GRITTING AND SALTING

During very cold weather ice can form on roads.
This is very dangerous.
To make roads safer they are gritted and salted.

Grit is spread to give tyres more grip.

Try this experiment.

board-slope covered with
1 smooth plastic, ice-like
   condition
2 sandpaper – gritted road

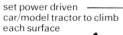

set power driven
car/model tractor to climb
each surface

thermometer

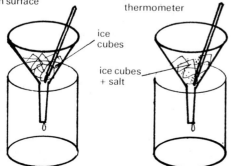

ice
cubes

ice cubes
+ salt

Why is salt spread on icy roads?
Try this experiment.

Observe and record what happens.

Time the melting.
Record the temperatures.

BRIDGING THE GAP

The bridge-builder has many problems.

The main one is that he has only short lengths to span a wide gap.

Your bridge-building will be with classroom materials.
Try bridging a wide gap with short pieces.
Use: newspaper rolls
     straws
     strips of wood
Can you bridge $\frac{1}{2}$m   1m   2m...?
Your investigations and answers are similar to those needed for real bridges.

SECTION THREE

A BRIDGE OF CARD

Cut a piece of thin card 60cm × 15cm.
Try bridging a 40cm gap.

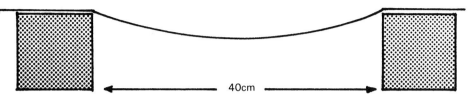

40cm

This does not make a very good bridge.
What can you do to it to make a better bridge?
Try folding and bending.

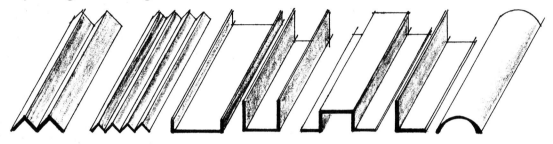

What loads will these carry before they collapse?

Find the strongest bridge that can be made from a 60cm × 15cm piece of card.

Remember to make your test fair. Each bridge must span the 40cm gap.

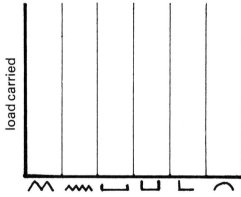

load carried

AND OTHER MATERIALS

Now try using different thicknesses and different kinds of material to bridge the gap.

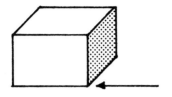 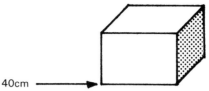

40cm

Try a 60cm × 15cm piece of these:
> sugar paper
> thin card
> thick card
> balsa wood
> corrugated card
> hardboard
> newspaper
> polystyrene

What load will each of these carry before it collapses?

Find the mass of each bridge.

Compare heaviness with strength.

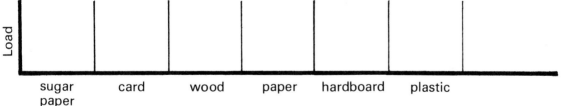

sugar paper	card	wood	paper	hardboard	plastic

Experiment by using more than one thickness.
Try gluing two or three thicknesses together.
Record the loads these bridges will hold up.
Draw a graph to show how the load carried changes as thicknesses increase.

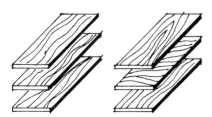

You could find out if sheets cut in different directions and glued hold up different loads.

Try 'sandwiches' of different materials.

BROAD AND NARROW

Early bridges were narrow.

They only needed to be wide enough for a single cart or pack-horse to cross them.

A narrow packhorse bridge

London Bridge

Modern bridges often have to carry several lanes of traffic.
How does width change the loads that can be carried?
Does the width of a bridge make a difference to its strength?

Experiment with different widths.
(Think about ways to make the test fair. Some parts of the test will have to stay the same each time.)

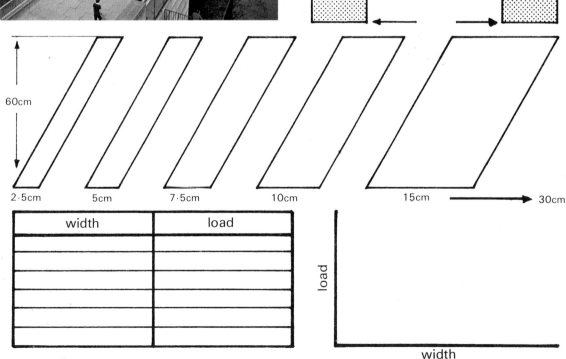

60cm

2·5cm   5cm   7·5cm   10cm   15cm   30cm

width	load

load

width

One of Britain's most famous bridges was built by Robert Stephenson.
This was the Britannia Tubular Bridge linking Wales and Anglesey.
Trains crossed the Menai Strait inside tubes.
Robert Stephenson's tubes were rectangular.
Is this the strongest shape?
Try making tubes of different shapes and test them.

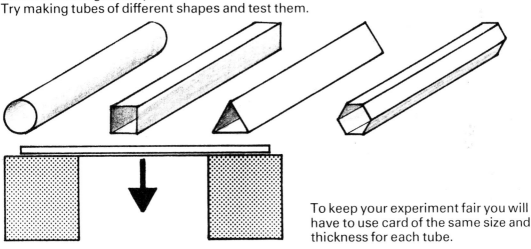

To keep your experiment fair you will have to use card of the same size and thickness for each tube.

What difference does diameter make?
Try tubes of different diameter.
Make your tubes by rolling paper round:
    thin dowel
    cane
    broom handle

Another test could be to make and test tubes of different kinds.

Try: newspaper
    sugar paper
    thin card
    more than one sheet

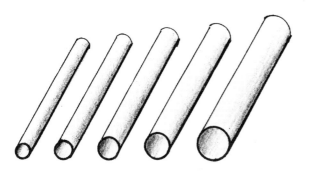

PILLARS AND PIERS

The Britannia Tubular Bridge was held up by five square-shaped pillars.

Many bridges have pillars like this supporting them.
Here is one at Arnside in Cumbria which has many such pillars.

Which shape of pillar makes strong bridge supports?

Use pieces of sugar paper 31cm × 10cm.
Make pillars of different shapes.

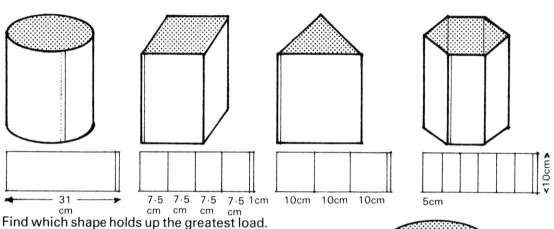

| 31 cm | 7·5 cm 7·5 cm 7·5 cm 7·5 cm 1cm | 10cm 10cm 10cm | 5cm | <10cm> |

Find which shape holds up the greatest load.

Make your experiment fair.
Use the same platform.
Carefully place your load in the centre.

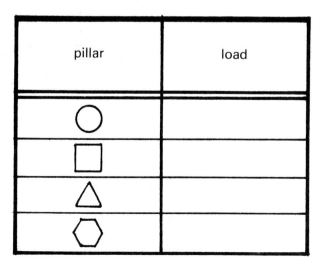

pillar	load
○	
□	
△	
⬡	

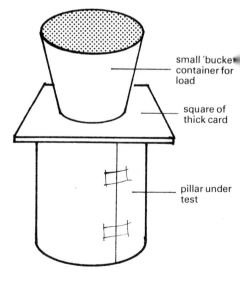

small 'bucket' container for load

square of thick card

pillar under test

PILLARS AND PIERS

What happens if the height of the pillar is changed?

Make pillars of different heights and load these.

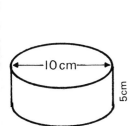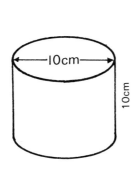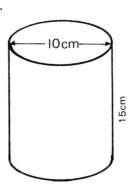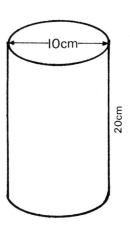

What happens if the diameter of the pillar is changed?
Make pillars of different diameters and load.

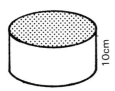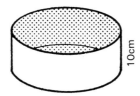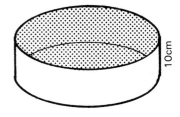

What happens if the load is put on unevenly?

Try this.

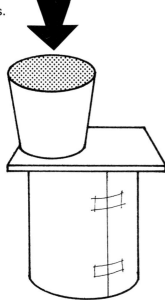

Some bridges stand in fast-flowing water. The shape of the pier then becomes very important. This bridge has boat-shapes. They are called cut-waters. Pages 21 and 22 of *Science in a Topic – Ships* suggest some experiments with these boat-shapes.

CROSSING THE BRIDGE

What happens as a load moves across a bridge?
Make a one-metre long bridge. Use scales for the piers.
Use pieces of polystyrene or balsa wood – light but rigid.

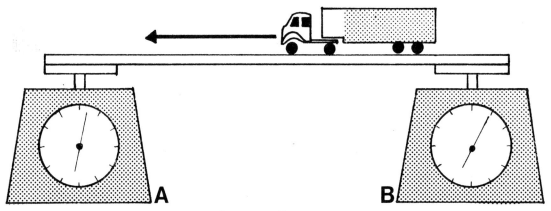

Move a load across the bridge, stopping every 5cm.
Record the two scale readings at each stop.

Distance stopped	Scale reading A	Scale reading B	A + B

If a very heavy lorry is crossing a bridge, where would be the most dangerous place for it to stop?

Work out a sum.

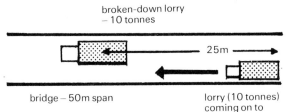

broken-down lorry
– 10 tonnes

25m

bridge – 50m span

lorry (10 tonnes) coming on to bridge

The breaking-point of the bridge is 15 tonnes.

Where will the second lorry get to before the bridge begins to break?

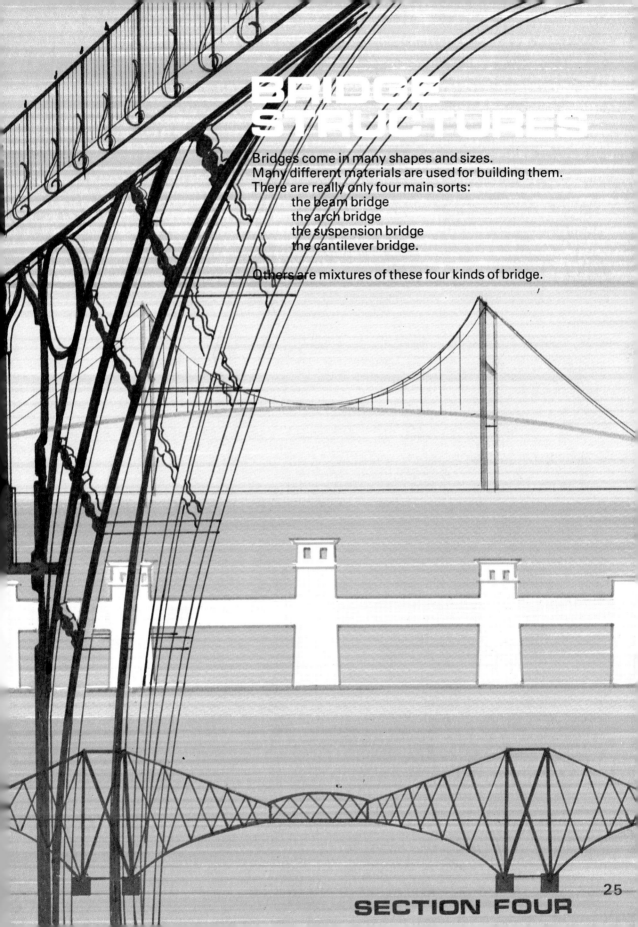

BRIDGE STRUCTURES

Bridges come in many shapes and sizes.
Many different materials are used for building them.
There are really only four main sorts:
- the beam bridge
- the arch bridge
- the suspension bridge
- the cantilever bridge.

Others are mixtures of these four kinds of bridge.

SECTION FOUR

THE BEAM BRIDGE

The very first bridge was probably a beam bridge.
Maybe a fallen tree bridged a stream.

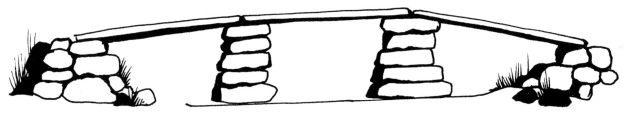

A plank between stepping stones would have helped the crossing.

Does it matter how the gap is bridged by the beam?
Experiment with a strip of balsa wood.
Find out if it is stronger on its edge.

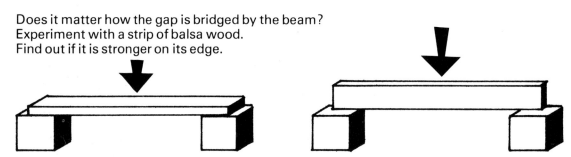

A bridge built from solid beams
could be very heavy.
It might even collapse under its
own weight.

Can you see from this picture how
the bridge-builder has overcome
this problem?

BEAM BRIDGES: BUILDING & TESTING

Is a constructed beam any weaker or stronger than a solid one?

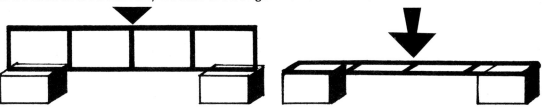

You could make other designs and test to find the strongest.
Here are some designs one group thought of.

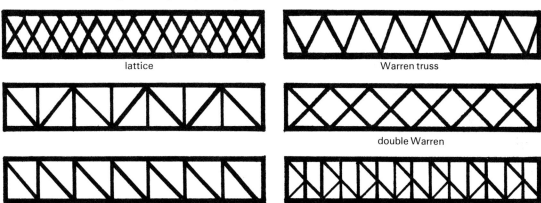

lattice

Warren truss

double Warren

'N' or 'Pratt' truss

'K' or Pennsylvania truss

Which do you think proved the weakest and strongest designs?
Here are some bridges using designs like these.

Newton's Mathematical bridge

Connel Ferry Bridge, Argyll

Windsor Rail Bridge

Rail bridge

STRENGTHENING BEAM BRIDGES

Find how and where beam type bridges can be strengthened.
Start with this bridge.

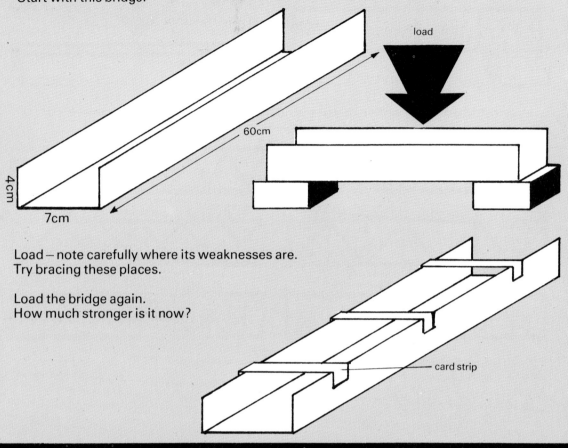

Load – note carefully where its weaknesses are.
Try bracing these places.

Load the bridge again.
How much stronger is it now?

The tubular bridge (page 21) is a type of beam bridge.
Try ways of strengthening a tube bridge.

Make balsa wood framing
to fit outside

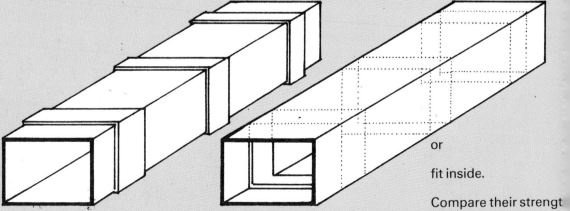

or

fit inside.

Compare their strengt

ARCHES

Arches have been used by bridge-builders since earliest times.

These arches were built by the Romans.

Roman Aqueduct, Caesarea

The arch has been used in many ways:

to carry the deck

to suspend the deck

to take the deck through

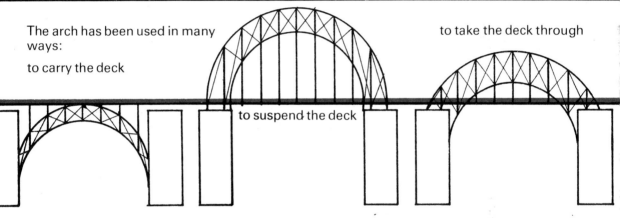

The arch has been built of many different materials.
As new materials have been discovered, bridge-builders have used them to build the arch.
Arch bridges have been built from stone, wood, iron, steel and concrete.

Here are some names used for parts of arch bridges.
Use your dictionary.
Match them to the picture.

> pier
> abutment
> spandrel
> voussoir
> springer
> keystone
> parapet
> deck

29

Clachan Bridge (The Bridge over the Atlantic)   Lambeth Bridge

An arch can be tall and narrow or wide and shallow.
How do the curves of arches change strengths?

Use a piece of card 40cm × 10cm.
Vary the distance between the abutments.
Load the arch.

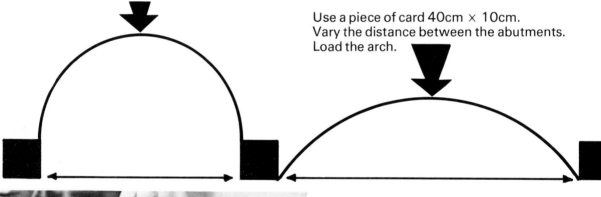

Record:

Distance	Height	Load

Which kind of arch is strongest?

HOW STRONG?

ndsor Rail Viaduct

Pont au Change, Paris

Another test on arches could be to keep the span distance the same. This time change the length of card making the arch.

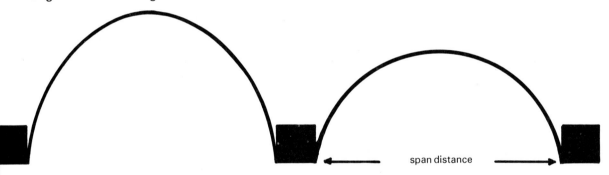

span distance

Record:

Length	Height of arch	Load

Does a mixture of beam and arch make a stronger bridge?
Load these.
Experiment with distances, heights and widths. Find the arrangement that makes the strongest deck arch bridge.

KEYSTONES AND FALSEWORK

Study the picture of this stone arch.
See how the arch stones are fitted together.
Notice the keystone.
Why is this piece so important?

Try building an arch bridge.
Thick polystyrene is a good material for such a model (the kind that is often used for packing).
In some ways it is like stone.
It is stronger in *compression* than in *tension*.

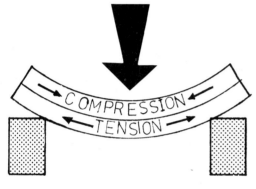

Draw the pattern of the stones on paper.
Transfer the drawing to your polystyrene.

Cut out the shapes.
Try building your arch.
How difficult is it?

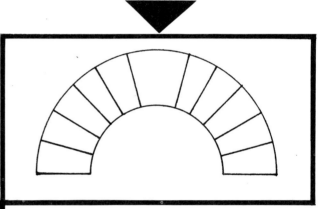

How do bridge-builders, working with heavy stone blocks, manage?

Falsework is built to hold the blocks in place.

Build some falsework for your arch.

See if you can make the curve using straight pieces of wood.
Strips of balsa wood have been used for this model.

SUSPENSION BRIDGES

Water has always been a main obstacle for the bridge-builder to span.
In warm lands, vines and creepers overhang the water.
These were the first 'suspension bridges', on which men could swing and clamber across.

Here are three famous suspension bridges.
What water does each of them span?

Severn Road Bridge

Bosporus Bridge

Spanning the water is often not the only problem. Ships may still need to pass. So a suspension bridge needs to be high enough. This problem has been solved in other ways with other sorts of bridge.
Collect pictures of bridges which open for ships to pass.

The Golden Gate Bridge

CABLE AND CHAIN

The best way to learn about suspension bridges is to build one yourself.

Build a model or work on a larger scale outdoors.

Here is a model which you can load. Find out about the pushes and pulls.

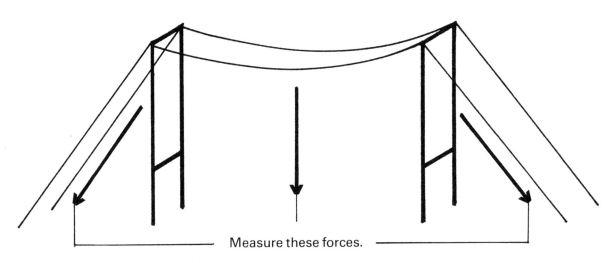

Measure these forces.

Here are some children working on a larger scale.
They are trying to find out if this angle is important.

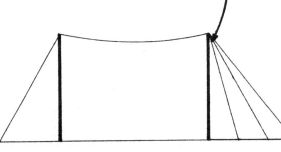

All suspension bridges have tall towers that hold the cables.
The ends of the cables have to be firmly anchored.
Here is the anchorage for the cables of the Humber Bridge.

The graceful curves of the suspension bridge cables make them one of man's most beautiful structures.

This is not done by chance.
Every measurement and force has to be carefully worked out.
Engineers plan and test and measure before they build.

Try some engineering drawing.

Make a scale drawing of a suspension bridge.

Use a scale of 1cm to 10m.
 The bridge is to be 300m long.
 The towers are to be 75m high.
 The deck hanging cables (hangers) are to be spaced 20m apart.
 The centre cable is to be 5m from the deck.

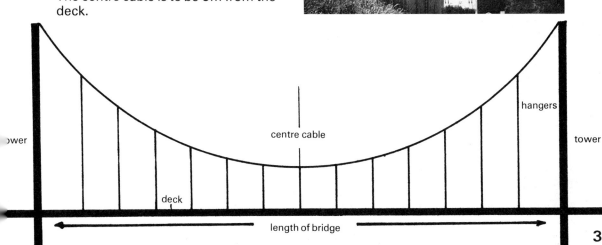

tower   centre cable   hangers   tower

deck

length of bridge

CABLE SPINNING

A problem when building a suspension bridge is to get the first line across the gap.

Some successful ways have been:

 by boat
 by kite
 by arrow
 by rocket

Once a thin link is across, this is used to haul thicker, stronger cables over.
Giant iron chains and thick cables are then used to suspend the deck.

John Roebling (1806–1869) used bundles of wire cables to build the world's first railway suspension bridge in the U.S.A.
Since then stranded wire cables have been used for all the world's great suspension bridges.
Spinning wheels go back and forth pulling two strands of wire. These build a giant cable that will hold the deck.
Each cable of the Humber Bridge needed 3 750 trips to make it. 36 000km (22 500 miles) of wire were used.
Each cable has 37 bundles of strands.
Why 37?
Remember the strands are round.

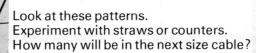

Look at these patterns.
Experiment with straws or counters.
How many will be in the next size cable?

RHYTHM AND BEAT-
SWAYING AND POUNDING

It is possible for a suspension bridge to sway, twist and collapse.
This can be caused by the wind. It can even be caused by a pattern of pounding feet.
Marching soldiers can do this.
The bridge can twist and shake to a regular beat. This movement grows until the bridge is torn apart.
This happened when a troop of soldiers marched across Broughton suspension bridge in 1829.

More recently a bridge in the U.S.A. twisted itself to destruction. In 1940 the Tacoma Narrows suspension bridge started swaying in a slight breeze. Within an hour the movement had reached a rhythm that twisted steel and concrete to destruction.

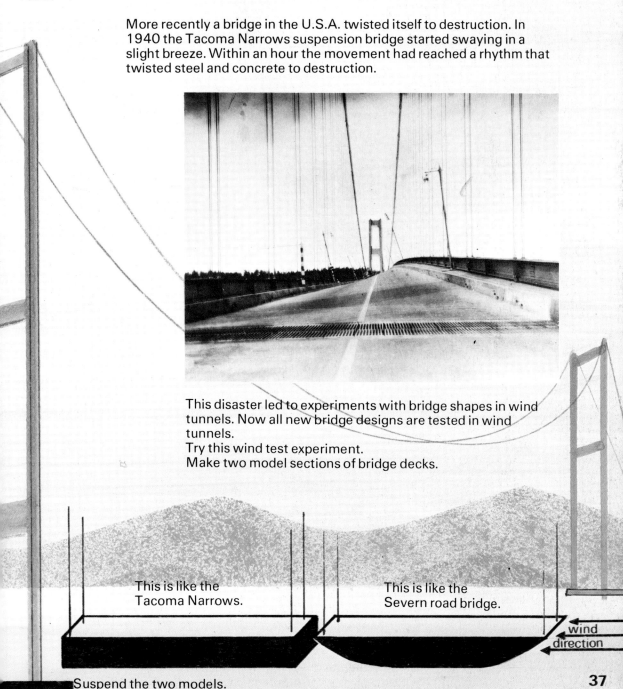

This disaster led to experiments with bridge shapes in wind tunnels. Now all new bridge designs are tested in wind tunnels.
Try this wind test experiment.
Make two model sections of bridge decks.

This is like the
Tacoma Narrows.

This is like the
Severn road bridge.

wind
direction

Suspend the two models.
Use a fan to test the sections.

CANTILEVER BRIDGES

This is the Forth Railway Bridge near Edinburgh in Scotland.

It is a cantilever bridge.
A cantilever bridge is one that has fixed arms. Sometimes it has a third section supported by these arms.

The forces in a cantilever bridge are complicated.
There are pushes and pulls in different directions.
Mr. Benjamin Baker who designed the Forth Bridge had trouble explaining these to the people of his time.
He did this by using men as the bridge parts.

Here are some children repeating this demonstration.

CANTILEVER FORCES

Experiment to find out about the forces in a cantilever bridge.
Find how a force in one direction must equal a force in the opposite direction.
If not, the bridge moves!
Build a simple cantilever bridge.

Place a load at the centre.
Do this slowly and carefully.
Notice what happens.
Where will you have to place loads to
balance the downward force of the centre load?

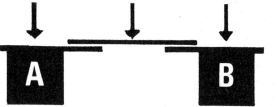

How big will they need to be?
Predict, that is make a guess or estimate,
then try.
Use one set of balanced loads that made
a good bridge.
Try different lengths of bridge pieces.
Now try different load positions.

Load at centre	Load on A		Load on B	
	Predicted	Real	Predicted	Real

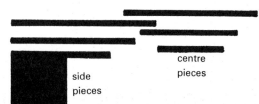

side
pieces

centre
pieces

CANTILEVER SHAPES

Try designing and building some cantilever bridges.
Here are some suggestions.

Draw your design on paper.
Cut it out.
Use this paper 'template' to cut four shapes from sheet balsa wood.
Join pairs of shapes with struts.
Cut sheet balsa wood for the connecting bridge decks.

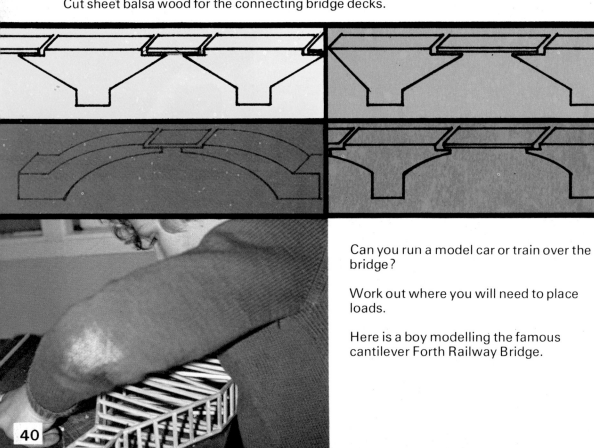

Can you run a model car or train over the bridge?

Work out where you will need to place loads.

Here is a boy modelling the famous cantilever Forth Railway Bridge.

TUNNELS

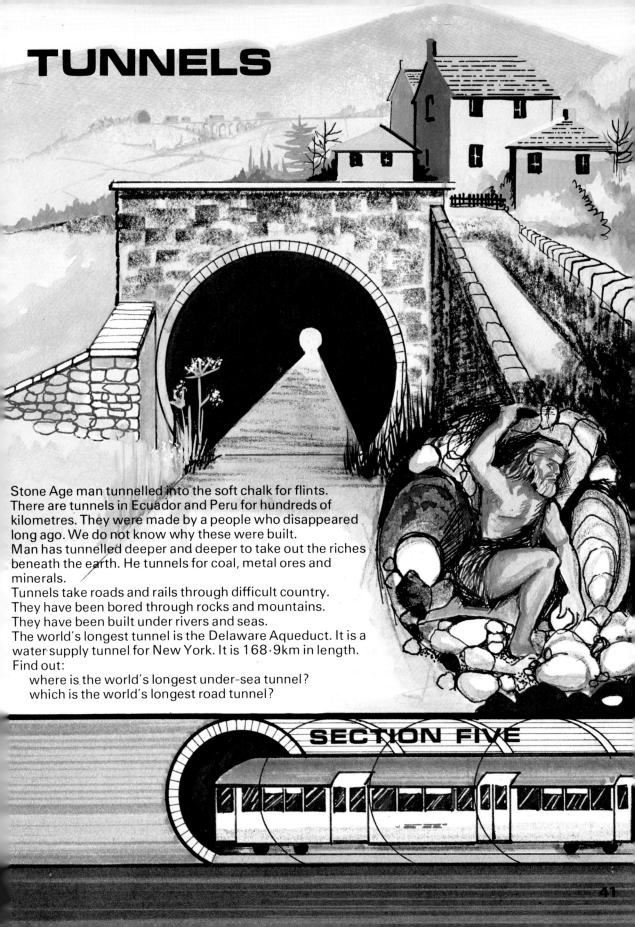

Stone Age man tunnelled into the soft chalk for flints.
There are tunnels in Ecuador and Peru for hundreds of
kilometres. They were made by a people who disappeared
long ago. We do not know why these were built.
Man has tunnelled deeper and deeper to take out the riches
beneath the earth. He tunnels for coal, metal ores and
minerals.
Tunnels take roads and rails through difficult country.
They have been bored through rocks and mountains.
They have been built under rivers and seas.
The world's longest tunnel is the Delaware Aqueduct. It is a
water supply tunnel for New York. It is 168·9km in length.
Find out:
   where is the world's longest under-sea tunnel?
   which is the world's longest road tunnel?

SECTION FIVE

TUNNELLING

There are three main ways that engineers use to build tunnels.

1 CUT AND COVER
  They dig a trench.
  They build the tunnel shape in the trench.
  The earth is put back on top.

2 THE SUNKEN TUBE
  This is used for making a tunnel under water.
  The tunnel shape is built on dry land. It is made of steel or concrete.
  A trench is dug in the river bed.
  The tunnel shapes are sunk into the trench.
  They are joined together under water.
  The trench is covered in.

Hong Kong Tunnel

Hong Kong Tunnel

3 THE DRIVEN TUNNEL
  This type of tunnel is bored through solid rock or earth.
  The men work inside the tunnel to do this.

You could use a sand pit to try tunnelling experiments for yourself.
(A sand play area or jumping pit would be ideal.)

DO NOT DIG PITS AND TUNNELS BIG ENOUGH FOR YOU TO GO INTO THEM.
THEY COULD COLLAPSE AND BURY YOU.

1 Card shapes are being tried for a cut-and-cover tunnel.

2 Open-ended tins are being tried as tunnel lining.

TUNNEL SHAPES

These are some of the shapes that have been used for tunnels.

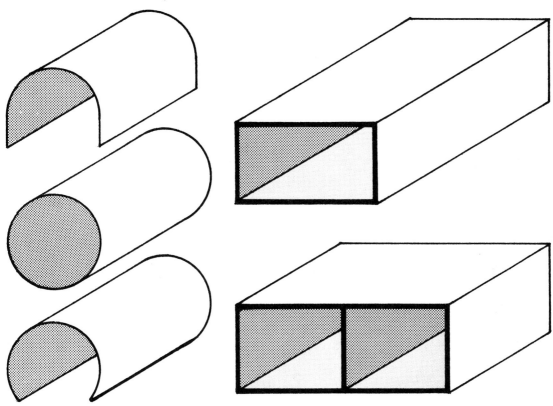

These tunnel shapes can have a mountain pressing down on them.
Which shape is the strongest?

Make these tunnel shapes from card.
Load them.

Record:

tunnel shape	load

7·5 cm

20cm

15cm

TUNNEL DESIGN

Many tunnels need to carry trucks and lorries.
Choose a model lorry. Design a card tunnel $\frac{1}{2}$m long that will take this truck.
Design another that will take traffic in both directions.
Design a third that will carry the trucks both ways, and also two rail tracks.

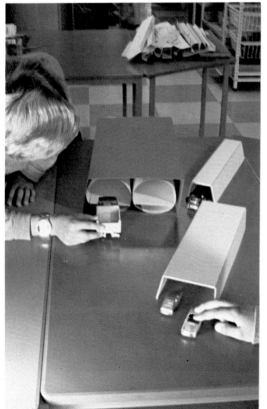

Test your designs to find:
1 strengths
2 which used least card
3 which shape needed to be strengthened?

What other things can be important in tunnel design?
Think about:
a accidents
b breakdowns
c fire
d air supply
e light
f water

GEOLOGY

n 1827 Brunel was tunnelling under the Thames.
Water rushed in, killed six men and injured Brunel.

In 1882, the St. Gotthard Tunnel under the Alps was completed.
Over a thousand men had been killed or seriously hurt during the building of the tunnel.
Other tunnellers have faced breaking into underground rivers, soft, waterlogged sand,
blistering hot rock and loose shifting gravel.

Today we know a lot more about the earth and its rocks.
This knowledge has made tunnelling easier and safer.
This science, the study of the earth's crust, is called *geology*.

Here is a geological collection made by a group of young geologists.

Make a rock collection yourself.
Use the collection to help answer these
questions.

1 Which specimens are rounded?
   Can you explain why?

2 Which specimens have a band
   of colour?
   Can you think out a reason for this?

3 Which specimens look like a
   'pudding' mixture?
   What explanation have you for this?

4 How many specimens have a
   'layered' structure?
   What other structures are there?

5 Do any specimens have holes
   in them?
   How did they get there?

Now take a specimen. List all you can say about it.
Can a friend pick your specimen from a group using your description?
Which descriptions were helpful?
Which were not at all helpful?

ROCK HARD

Geologists have tests to find out about rocks.
The hardness of rocks is important to tunnellers.
The geologist's test for hardness is a 'scratch test'.

In this list each rock or mineral will scratch the one below it.
The list is called 'Mohs' Scale of Hardness'.
(Friedrich Mohs was a German geologist:)

MOHS' SCALE	
10 Diamond	
9 Corundum	
8 Topaz	
7 Quartz	(a hard steel file will scratch)
6 Orthoclase feldspar	
5 Apatite	(can be scratched with a steel penknife)
4 Fluorite	(can be scratched by glass)
3 Calcite	(can be scratched by an iron nail)
2 Gypsum	(can be scratched with the finger nail)
1 Talc	(can be crushed with the finger nail)

Diamond is the hardest. Talc is the softest.
Which rocks on the scale will quartz scratch?
Which will it not scratch?
If you test a rock and fluorite scratches it, what test do you try next?

The hardness of rock decides which method of tunnel drilling is used.

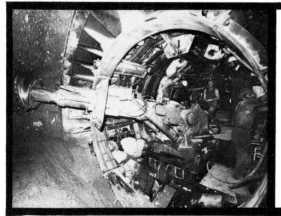

For hardness 1 to 5 a turning cutter type drill is used.

For hardness 3 to 7 a hammer type drill is needed.

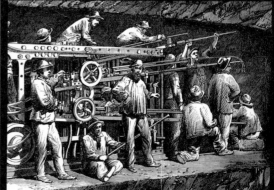

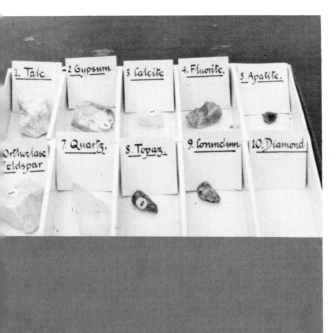

ROCK HARD

A collection of rocks graded as Mohs' scale can be bought.

Everyday things can be used to test your rocks and stones.

A steel file has a hardness of about 6.5.
It will scratch all rocks softer than this.

A penknife will scratch rock 5.5 and below.

A copper coin has a hardness of about 4.

Your finger nail is about 2.5.

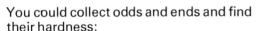

You could collect odds and ends and find their hardness:

> a piece of flint
> a masonry nail
> a woodwork nail
> a piece of plastic
> a piece of wood
> a drawing pin
> a paper clip
> a piece of broken pottery

Now try scratch tests on your rock collection. Grade your samples in order of hardness.

What would the drill cutter need to be made of to cut through the hardest rock in your collection?

TUNNELS AND WATER

Water has always been a main enemy of tunnellers.
They sometimes meet underground springs and streams.
Water can burst into the tunnel or seep through the rocks.
Geologists measure how much water rocks can soak up and let through.

See how classroom chalk soaks up water.

Obtain a lump of chalk.
Weigh it.
Soak it for five minutes.
Weigh it.
Repeat this until there is no change.
See how long it takes for the water
to dry out.
A graph would be a good way to show
these changes.

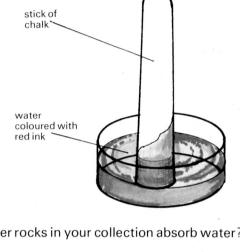

stick of
chalk

water
coloured with
red ink

Which other rocks in your collection absorb water?

See how water seeps through different rocks and earths.

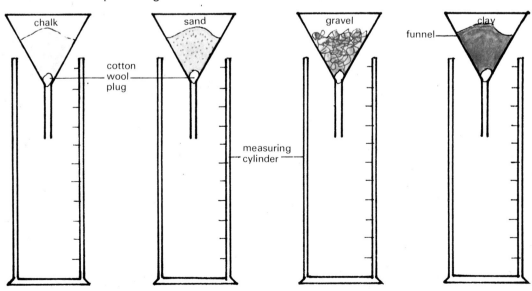

Use 100ml of water and observe how fast the water seeps through.

THE BUILDING MATERIALS

Many different materials have been used to
make roads, bridges and tunnels.
First, they were the materials to hand:
    stones, clay and mud
    vines from the forest
    fallen trees and branches

New tools and new ideas helped to build
better structures:
    stone blocks could be cut and carved
    wood was shaped and jointed
    metals were mixed and cast

Now we have made new materials:
    cast iron
    wrought iron
    steel
    concrete
    reinforced concrete
These mean that we can build better
and bigger structures.

SECTION SIX

WOOD – HOW STRONG?

Wood has always been one of man's most important materials.
Wood blocks were used for roads.
Bridges were built of wood.
Tunnellers still prefer wooden props. (They give warning creaks.)

How strong is wood?
(Lolly sticks or balsa wood strips are good samples to test.)
Here is a group testing the strength of a wood strip.

DO BE CAREFUL.

KEEP TOES AND FINGERS CL[...]

AS WOOD BREAKS IT MAY
JERK UP AND SPLINTER.

A newspaper or cloth pad will protect the floor.

Measure the breaking force for
different lengths.

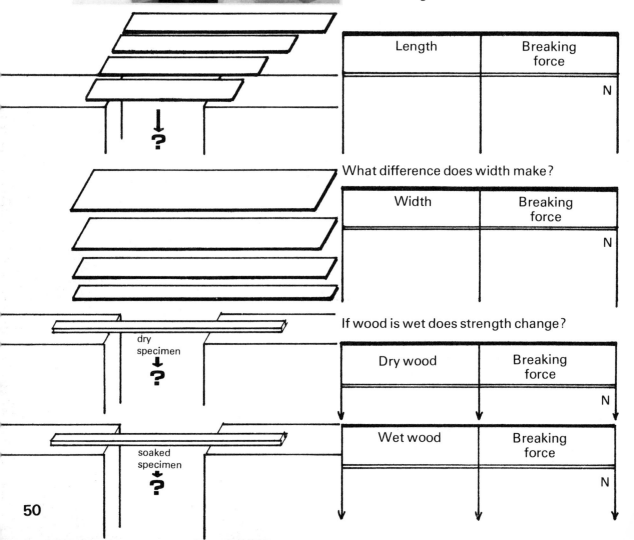

Length	Breaking force
	N

What difference does width make?

Width	Breaking force
	N

If wood is wet does strength change?

Dry wood	Breaking force
	N

Wet wood	Breaking force
	N

WOOD – HOW BENDY?

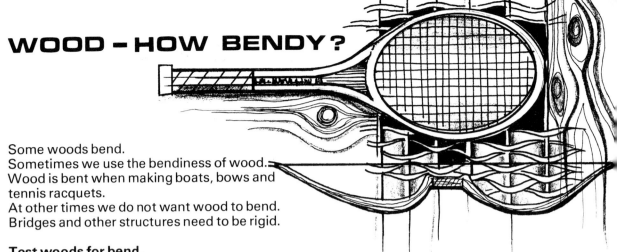

Some woods bend.
Sometimes we use the bendiness of wood.
Wood is bent when making boats, bows and
tennis racquets.
At other times we do not want wood to bend.
Bridges and other structures need to be rigid.

Test woods for bend.

Make a chart to show how bend
alters with load.

Experiment with different woods,
different thicknesses, different
lengths and different widths.

Wood under test	Force load	Bending noted

Try wetting samples to see if bending changes.

Bridge-builders need to know how much
materials bend.
They then know which materials to use and how
long and thick the piece needs to be.

WOOD-ROTTING & PRESERVING

Wood rots.
Wood used for bridges and structures
must be protected.

These are some of the things
used to protect wood.

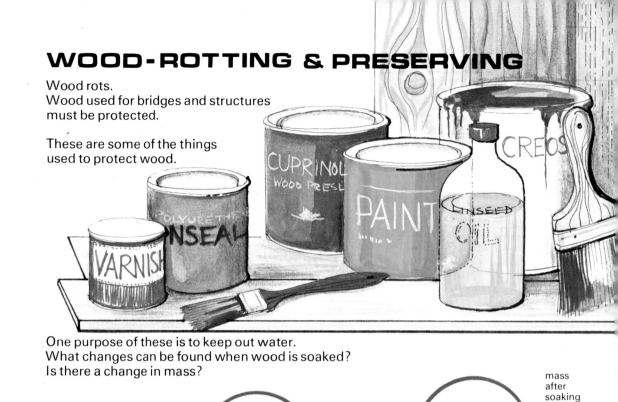

One purpose of these is to keep out water.
What changes can be found when wood is soaked?
Is there a change in mass?

Is there a change in size?

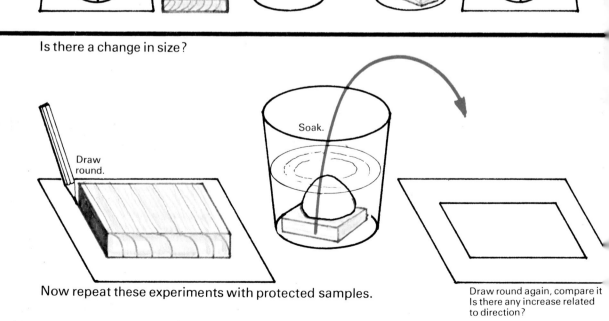

Now repeat these experiments with protected samples.

METALS - THE IRON BRIDGE

This is Iron Bridge.
It spans the River Severn at Coalbrookdale in Shropshire.
It is famous because it was the first bridge in the world to be made of iron.
Coalbrookdale was the centre of the iron industry in the eighteenth century.
Iron ore, limestone, fuel, firebricks and the resulting pig iron had to be ferried to and fro across the river.

In 1775 it was decided to build a bridge. It was also decided to build from the new material, iron.

Abraham Darby, master of an ironworks, was in charge of the building.

The pieces of the bridge were cast. This means that molten iron was poured into moulds.

There was no knowledge of how to join such large pieces of iron. It is interesting to see that it was fastened in the way wood would be fastened.

Traffic no longer crosses the bridge. It is a National Monument.

METALS-SPRING AND BEND

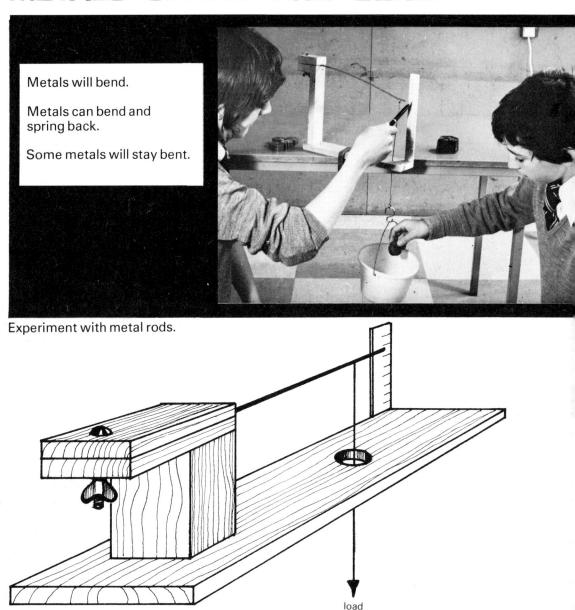

Metals will bend.

Metals can bend and
spring back.

Some metals will stay bent.

Experiment with metal rods.

load

Record what happens as the rod is loaded step by step.
Now unload following the same pattern, and record bend.

Metal	Load	Bending	Unload	Bending

METALS - HOTTING UP

Find out what happens when metals are heated.
One thing is sure – they get hot. Be careful.

These pieces of equipment will show that metals *expand* when hot.

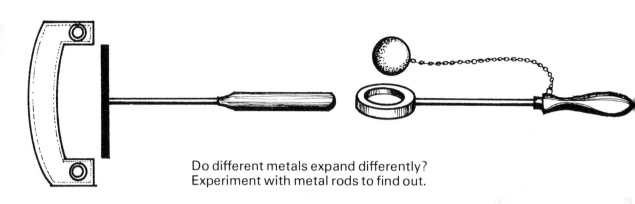

Do different metals expand differently?
Experiment with metal rods to find out.

Small changes are
difficult to see.

Here is one school's way
of measuring this.

Expansion of metals can cause problems.
Gaps are left when using metals to build
roads, bridges and tunnels.
Here are some gaps that allow for expansion

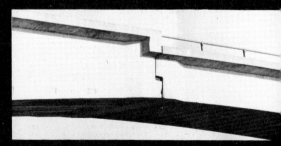

Expansion has also been used to help builders.
The Forth Rail Bridge was 20cm (4in) too short. A fire was lighted, the piece
expanded and the fastening was made.

CONCRETE

It is known that the Romans used a kind of concrete.

It was Joseph Aspden who developed modern concrete.
In 1824 he burned limestone and clay in his kitchen stove. This is cement. When mixed with water a *chemical reaction* happens and it sets hard.

When a mixture of gravel, sand, cement and water is used, this is concrete.

The use of this material has led to the design of some of the world's most graceful yet strong bridges.

Motorway Bridge, M40

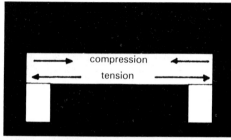

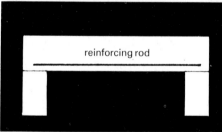

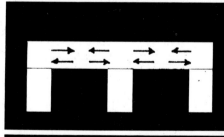

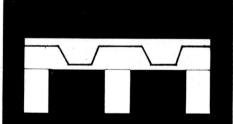

Concrete is very strong when squashed (in *compression*).

It is weak when pulled (in *tension*).

In 1880 there was an idea to strengthen concrete. Iron bars were added. This became known as reinforced concrete.
Experiment for yourself and find out about reinforced concrete.

Make a mould.

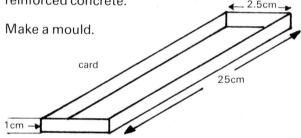

Mix some concrete. (One part cement, four parts sand and stones, water to mix — not too wet.)

Try reinforcing with 1 rod — 2 rods — 3 rods — 4 rods — no rods. (Florist's wire will make good rods.)

Load your samples and compare strengths.

BE CAREFUL OF TOES — quite large loads will be needed to break some samples.

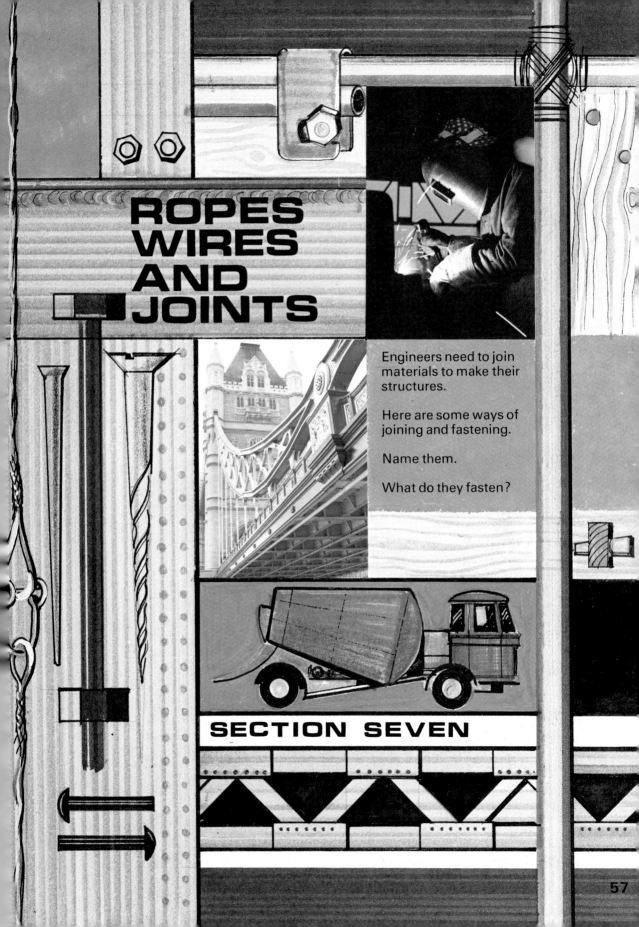

ROPES WIRES AND JOINTS

Engineers need to join materials to make their structures.

Here are some ways of joining and fastening.

Name them.

What do they fasten?

SECTION SEVEN

57

STRAIN AND STRETCH

Very large forces are at work in a bridge like this.

Engineers need to know the strengths of the wires.

They also need to know how much they will stretch.

The Albert Bridge

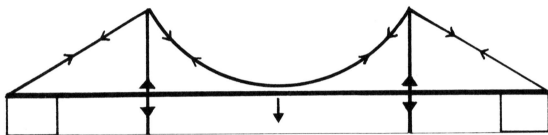

Measure the strength and the stretch of different wires.

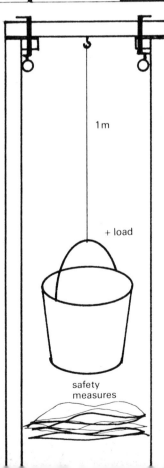

wire	stretch	breaking force
	mm	N
	mm	N

1m

+ load

safety measures

Try: fuse wires
      thin copper wire (26 – 34 s.w.g.)
      constantan wire   „
      nickel chrome wire   „

WEAR SAFETY GOGGLES AND GLOVES –
BREAKING WIRE CAN BE DANGEROUS.

Compare wire strengths with threads.
Try: cotton
      nylon (fishing line)
      silk
      wool

MORE THAN ONE – WIRES AND ROPES

The bridge-builder uses many wire strands in his cables.
All cables and ropes use many strands.
They can be made in different ways.

Try twisting and plaiting some threads and comparing strengths.
Try: 1 thread
       2 threads (looped)

       2 threads (twisted)

       3 threads (twisted)

       3 threads (plaited)

Use your test rig or there is a 'thread breaker' in *Science in a Topic – Clothes and Costume*, page 19.

When ropes are used they have to be tied and lashed.
A knowledge of knots is needed.
Who uses these knots?
Practise tying and using them.

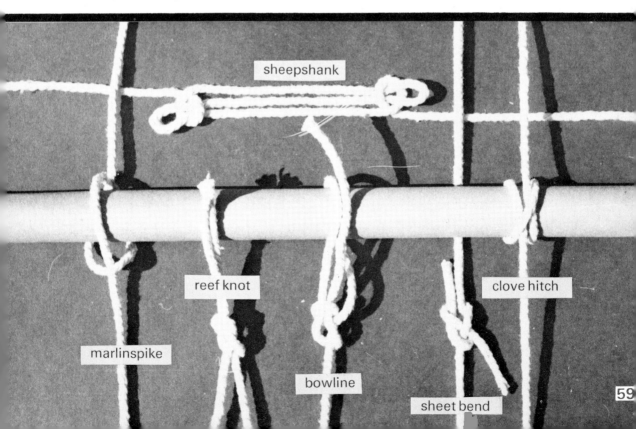

sheepshank

reef knot

clove hitch

marlinspike

bowline

sheet bend

MORE THAN ONE
ROPES & WIRES

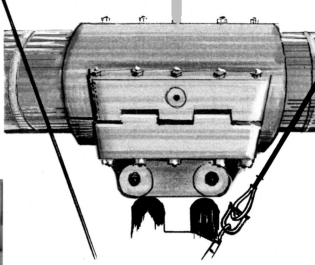

To gain strength builders use more than
one piece.
Are two pieces twice as strong as one?
They may be more or less than twice the
strength.

Experiment with: straws
                    plant stems
                    strips of balsa wood
Try 2 – 3 – 4 – or 5 in a bundle.
(Secure with sticky tape or thread.)

Measure the force needed to break the bundle.

Breaking force				
1	2	3	4	5
N	N	N	N	N

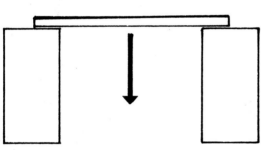

Is there a strength pattern?

Are there differences if the bundles are
arranged differently?

Try bundles of three.

Try bigger bundles.

Is the strength pattern the same for different
materials?

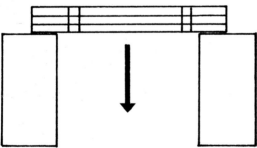

Here are some ways used in joining wood.

Here are ways in which metal is joined.

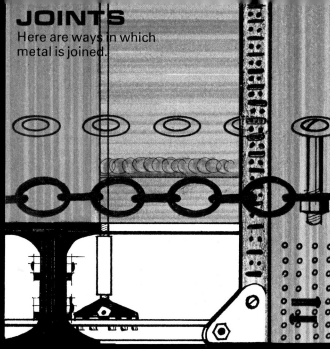

Stones and bricks are joined by mortar.

The way in which strips and beams are fixed can make all the difference to strength.
Try fixing strips together to make shapes.

Try 3, 4, 5, 6, 7 . . . strips.

Which are rigid shapes?
What can you do to make the non-rigid shapes rigid?

Engineers test and use strong materials.
They need to join them.
The joints are often the weakest parts.

61

GLUES

You will have used glue to make models.

Of course, real bridges would not be fastened together with glue. However, a bridge has been built of paper and glue that carried a lorry across a gap.

Which is the best glue?

What do we mean by best?

Is it: the fastest drying?

the one that will stick the greatest number of different materials?

the one that makes the strongest joint?

You could test for each of these.

The test to find the strongest would be most useful.

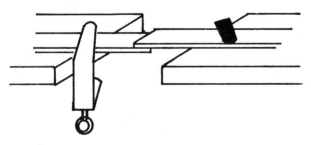

Use wooden lolly sticks or hardboard strips.

Think about:
(a) how much glue you use
(b) the area you cover with glue
(c) the surfaces you glue, rough or smooth
(d) the time left before testing

slight overlap   butt joint

Test some of your glue joints after soaking in water.

One group of glue-testers found their joints were very weak.
They discovered why it was important to read the maker's instructions!

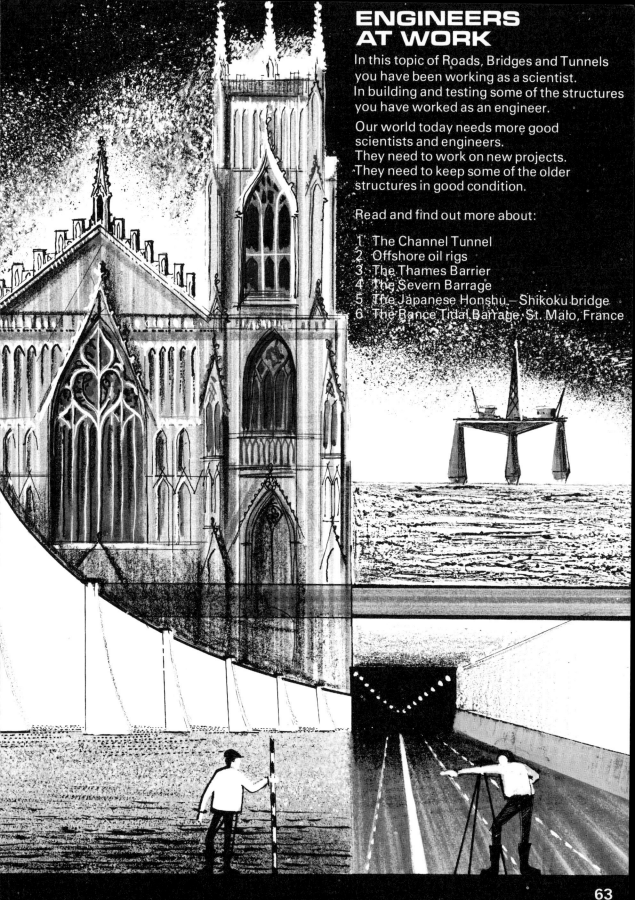

ENGINEERS AT WORK

In this topic of Roads, Bridges and Tunnels you have been working as a scientist.
In building and testing some of the structures you have worked as an engineer.

Our world today needs more good scientists and engineers.
They need to work on new projects.
They need to keep some of the older structures in good condition.

Read and find out more about:

1. The Channel Tunnel
2. Offshore oil rigs
3. The Thames Barrier
4. The Severn Barrage
5. The Japanese Honshu.—Shikoku bridge
6. The Rance Tidal Barrage, St. Malo, France

MATHEMATICS	ART AND CRAFT	ENGLISH

Surveying – Measurement
Levels, Slopes, Cambers
Number Patterns
Tests, Records – Graphs,
   Charts
Models – Scales
Measurement of Forces
Ratios – Relationships
Records, Spans – Lengths,
   Heights
Rigidity of Shapes
	Model-Making
	Test Rigs – Technology
	Landscapes – Layouts
	Paper Sculpture
	Collage
	Thread Pictures
	Plans – Cross Sections
	Working in Wood and Metal
	Working with Stone and Plastic
	Motorway Photo Montage*
		Road Themes in Legends,
		   Ballads, Rhymes and Sayings
		Chesterton: *The Rolling
		   English Road*
		Longfellow: *Paul Revere's Ride*
		Wordsworth: *Lucy Grey*
		Colum: *An Old Woman of the
		   Roads*
		Stevenson: *The Vagabond*
		Milne: The Game of Pooh
		   Sticks, in *The House at
		   Pooh Corner*
		Williams: *The Wooden Horse*
		Place Name Studies
		Biographies of Great Engineers
		Creative Writing – Travel and
		   Adventure

ROADS, BRIDGES AND TUNNELS

HISTORY	GEOGRAPHY	R.E. AND NATURAL HISTORY

Ancient Trackways
Roman Roads
Pack-horse and Footpads
Wade, Macadam and Metcalfe
Telford – Viaduct and Canal
Stage Coaches and
   Highwaymen
Toll Gates and Turnpikes
Abraham Darby and Iron
Brunel and The Great Western
   Railway
Bessemer and Steel
Tay Bridge Disaster
Cobbles, Stones, Setts and
   Concrete
Motorways – Planning and
   Protest
	Distance and Direction
	Geology – Types of Terrain
	Natural Barriers – Mountains
	   and Rivers
	Routes, Maps, Map-Reading
	Road Classification
	Siting and Growth of Towns
	Great Roads – Great Routes
	Finding the Routes – The
	   Explorers
	Mineral Resources
	Asphalt Lake – Trinidad
	Trans-Continental Routes
	Roads and the Environment
		Journeys of the Israelites
		Parable of the Good Samaritan
		Paul and the Road to
		   Damascus
		The Road to Calvary
		The Road to Emmaus
		Bunyan: *Pilgrim's Progress*
		Crusades
		Pilgrimages and Holy Places
		
		Animal Burrowers and
		   Tunnellers:
		Rabbits and Other Rodents
		Otters, Badgers
		Snakes, Lizards, Ants

ACKNOWLEDGEMENTS

The authors would like to express their thanks to the following for help in supplying copyright photographs:
Mr. C. Jeavons; Mrs. Moore; the Swiss National Tourist Office and Swiss Federal Railways; the French Government Tourist Office; the Mansell Collection; Radio Times Hulton Picture Library; John Laing Construction Ltd.; Mr. W. Cardin, Highways Department, Buckinghamshire County Council; the Institution of Civil Engineers; Ealing Beck Ltd.; the Scottish Tourist Board; The Illustrated London News; Richard Costain Ltd.; Cippenham Casting Co. Ltd.; Mr. Cleland Rimmer; Ciba-Geigy (U.K.) Ltd.